"The Quiet 707 program made a major impact on Boeing, airlines, and the aviation industry as a whole. Your unique insight in to the program is a boon for researchers, and I am sure your book will be referenced for many years to come."

> Meredith Lowe
> Head Librarian
> The Museum of Flight

"The Quiet 707 program made a major impact on Boeing, airlines, and the aviation industry as a whole. Your unique insight in to the program is a boon for researchers, and I am sure your book will be referenced for many years to come."

>	Meredith Lowe
>	Head Librarian
>	The Museum of Flight

Quiet 707

Jack Shannon

First Edition
Copyright © 2013 by Malcus, Inc.
Jim Larsen photographs
All rights reserved
ISBN 978-1-300-63755-4

*This work is dedicated
to all those technically and scientifically
inclined individuals
who so want to start their own company,
but, by virtue of their perceptions and expertise,
are ill prepared for the difficulties that lie ahead.*

The Quiet 707

In 1982 a small and relatively unknown engineering group embarked on a path that gained international attention and altered the history of commercial aviation. In the next eight years they would complete the design, engineering, parts fabrication and testing to secure federal aviation and international approval for a "hush kit" on the fabled Boeing 707. As I will show later, that work would conflict with the express expectations of congress, the Federal Aviation Agency, NASA, the Boeing Company, and Mc Donald Douglas. But it would also produce several hundred million dollars in revenue for two large-scale aerospace companies and deliver more hushed airplanes for decades than the combined total of well over ten competitors. As the senior individual in that company I would learn about raising research and development capital and I would learn more than I cared to know about the unethical side of business, and become, for at least a short period, internationally famous in the industry.

This is a true story.

Acknowledgement

I am indebted to Dorianne Berry for contributing editing skills and in critiquing this work. Her experience as a radio and TV personality in South Africa lent a perspective not normally available to an engineer entrepreneur. I have found it incredibly tedious to succinctly organize one's recollections into a smooth narrative. The fact it took this many years to do so is evidence of that. The fact that Dorianne had absolutely no acquaintance with this technology or market gave her invaluable insight, promising to make this work more appealing to a much broader audience.

Her appearance at this time was synergistic and common to experiences related herein. It is those we find ourselves in contact with that make the journey so rewarding.

Acknowledgement

I am indebted to Dorianne Berry for contributing editing skills and in critiquing this work. Her experience as a radio and TV personality in South Africa lent a perspective not normally available to an engineer entrepreneur. I have found it incredibly tedious to succinctly organize one's recollections into a smooth narrative. The fact it took this many years to do so is evidence of that. The fact that Dorianne had absolutely no acquaintance with this technology or market gave her invaluable insight, promising to make this work more appealing to a much broader audience.

Her appearance at this time was synergistic and common to experiences related herein. It is those we find ourselves in contact with that make the journey so rewarding.

Table of Contents

Preamble		1
I	The Environment Leading to the Noise Rules	3
II	The Unique Nature of the 707 and DC-8	7
III	Studies by Boeing, NASA, the FAA and McDonald Douglas	9
IV	Company Background	11
V	The First Step	17
VI	Preliminary Configuration	21
VII	The Market for Venture Capital	31
VIII	Seeking Research and Development Capital	33
IX	Contract Negotiations	39
X	Contract Signing	43
XI	Development	47
XII	Flight Test	53
XIII	Certification	61
XIV	Business Dealings with Boeing	65
XV	Production	67
XVI	International Aircraft	73
XVII	Community Transports, Ltd., AKA Com Tran, Ltd.	79
XVIII	Post Flight	83
References		87

Illustrations

World Map of Countries with Noise Regulations	5
The Unique Nature of the Boeing 707 and DC-8	8
The Quiet 707 Nacelle Configuration	25
Reduced Landing Flap Switch	26
Quiet 707 Engine and Nacelle on the Test Stand	50
Test Crew in the Cockpit	57
Test Aircraft over Measurement Station at Goodyear	59
Warning Communication from Boeing	66
Customer Letter, Civil Aviation Department, Hong Kong	76
Customer Letter, Pakistan International	77
Customer Letter, United Arab Emirates	78

Preamble

This is a true story of a small engineering company that briefly made history. It's a story about the company founder who while skilled technically, could not comprehend the business aspects until well after the program was over.

I am that founder.

Now, almost thirty years after the endeavor, I realize that the benefit was in the experience. It was unique because of the international scope of the program; the wide variety of customers, some clearly involved in reprehensible, if not illegal businesses; and the dishonest street-tactics of some of the companies with whom I was to be involved.

This is also a story of how important luck is, and how events in the nation's tax structure, acts of congress in attempting to "help," and outright attempts at sabotage all play important and unpredictable roles. This is a story about how in retrospect business plans at best are naive and even comical.

It is also a story of how personal relationships and trust were the indispensible qualities. Without those qualities there would not have been a program, or this story.

I The Environment Leading to the Noise Rules

In 1969 congress proposed adding to the Code of Federal Regulations (CFR), Part 36, a section which established noise rules for new production aircraft. [A] The 707 and DC-8 were by then fifteen years old. Both were equipped with the JT3D-3B engine, a relatively noisy low by-pass ratio turbofan engine. By that year Boeing had delivered 1,500 727's, the 737 production line had been open for several years, and the DC-9 had been in production for four years. The 727, 737 and DC-9 were equipped with the JT8D, which had a higher by-pass ratio turbofan and a much quieter engine. Even in the mid 60's virtually every engine manufacturer had in development or production much quieter higher by-pass ratio turbofan engines. [B]

As a result of this proposed rule, considerable aircraft noise research was accomplished by: Boeing; the McDonald Douglas Corporation; the Federal Aviation Agency, FAA; and the National Aerospace and Space Administration, NASA, in the 1970's. This research was to determine whether it was feasible to reduce the noise of the DC-8 and 707 aircraft. The results of those studies and eventual flight tests determined it *might* be possible to do so. [C] The increase in direct operating costs was predicted to be between four and nine percent. [D] At that time the 707 fleet size was estimated to be 496 aircraft. *If* adequate program financing was available, the sell price was estimated at $760.000 per airplane. ^E A requirement of this research was that the final design be *"capable* of being certified to the applicable provisions of" the relevant airworthy regulations. That requirement was met by a Boeing Designated Engineering Representative of the FAA, a DER, signing a "Statement of Compliance with the Federal Aviation Regulations" that the design was " Flight worthy ... for the *intended* experimental Flight Test Program" ^E (Italics have been added for emphasis). In later years competitors to our program would argue that the design documented by these reports was *"certified"*, and the competitor need only build that design and sell those hush kits. The words in italics are self-serving and they created a large misunderstanding (see Chapters IX and XV).

But the real issue was whether newer, more efficient and quieter airplanes would be available by 1985. Wouldn't those airplanes be a better solution for the manufacturers and the general public? There was no doubt that these newer engines offered much better fuel economy, lower direct operating costs and they were going to be available by 1985. It was accepted by the industry that there would be no hush-kitted fifteen-year-old airplanes in any domestic service by that time. Hence, there was no need to contemplate a hush kit program on any of these airplanes.

By 1985 and 1988 the world-wide impact of these pending federal rules [F] and the International Civil Aviation Authority, ICAO, rules [G] are shown in this figure.

World Map of Countries with Noise Regulations on January 1985

And by January 1988

February 23, 1982 Briefing Illustrations (chapter V).

In 1976 the Federal Aviation Noise Abatement policy [F] was passed into law giving older, noisier airplanes such as the 707 and DC-8 until January 1 of 1985 to comply with Federal Aviation Regulations, FAR, 36 Stage 2 standards or be phased out.

By July of 1983 the Port Authority of New York and New Jersey responding to increasing public pressure, invoked noise restrictions that would prohibit operations by Boeing 707 and DC-8 aircraft.

The perception was "there was no quick or cheap way to make the planes conform with the noise restrictions." [H]

II The Unique Nature of the Boeing 707 and DC-8

But there was another factor at work. Even in the 1960's manufacturers had plans underway to produce airplanes larger than any before: the 747, DC-10, and L1011. The future market was perceived to be that of serving major cities. There, a passenger would board the short haul 727, 737, or DC-9 to continue to their destination. That left the 707 and DC-8 as the only "long thin" airplanes for the decade beyond the eighties.

The following figure from the February 23, 1982 briefing of 707 and DC-8 operators (Chapter V) shows the impact of the Noise Abatement policy. Those long - range thin aircraft, the 707 and DC-8, would be scrapped in 1985 by the enforcement of noise abatement policy.

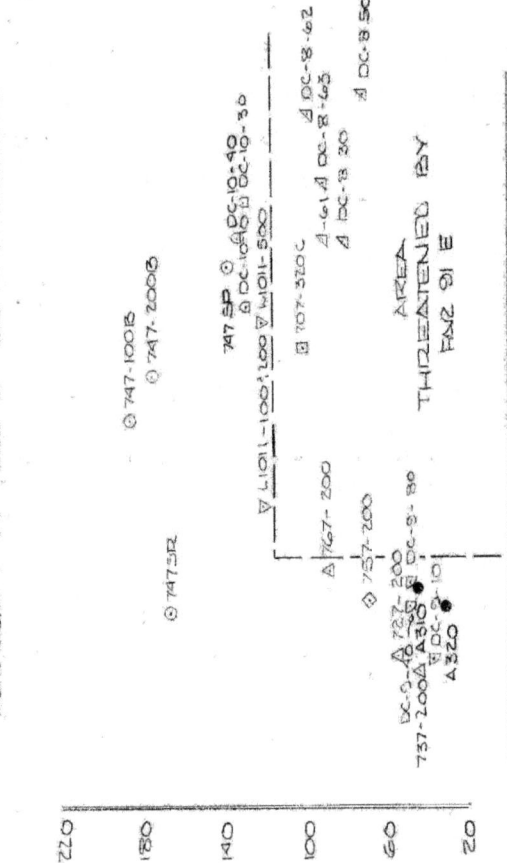

III Studies by Boeing, NASA, the FAA and McDonald Douglas

Before the noise rules were to be enforced, Boeing, NASA, the FAA and McDonald Douglas set about proving that the rules would not unduly encumber anyone currently operating 707's or DC-8's. That work consisted of fabricating nacelle modifications on a Boeing 707 and completing such flight tests as seemed appropriate to demonstrate compliance with the rules. The nacelle modifications consisted of a three quarter (relative to the length of the nacelle) fan-duct extension and acoustically treated rings in the inlet. The work, which included extensive data, was compiled in four volumes titled "FAA JT3D Quiet Nacelle Retrofit Feasibility Program." [A] To confirm the suitability of the design, a Boeing Designated Engineering Representative, DER, of the FAA completed an 8110 form stating that the design was "capable of being certified." [B] In fact the design had two serious shortcomings: the inlet had acoustically treated rings which were not anti-iced and no means existed to achieve anti-icing; and the fan duct extension was such that sonic-flow occurred in cruise, between that surface and the lower wing, to cause an unacceptable drag rise. The DER's statement was true in that the design was "flight worthy ... for the *intended* experimental Flight Test Program," and the configuration, or more correctly *something like that*, could be certified, but far more work had to be accomplished.

The DER's statement, while well intended, did appear to demonstrate that the design goal had been met, but it proved to be a major problem as 1985 approached. At this point numerous groups entered into contracts to build this specific design claiming that it had already been certified.

IV Company Background

My background was in engineering flight testing. And so, I typically utilized pre-flight and post-flight meetings. Just as one would pre-flight an airplane; we would discuss the objectives: our strengths, areas of risk, and options if necessary. This is a chapter about our pre-flight status.

I had spent two summers prior to graduation at Patuxent Naval Air Test Center in Maryland. My graduation was in 1961 with a degree in Aero Space Engineering from The University of Texas.

I was then hired by Boeing Seattle for their flight test group. I left Boeing around 1970. Before I left Boeing I had assignments in acoustics, and fighter and supersonic airplane design. For me, my Boeing experience was as if I was in a graduate program at the most prestigious university in the world. I could walk the halls at Boeing, talk to real experts and learn more than I could ever remember about any technical subject as long as it related to transport airplane design, manufacture or operations.

One of the things I learned when I was assigned to the Boeing flight test group was the benefit of an orderly documentation system. An intelligent perspective is to know how much it would cost to arrive at a conclusion or to glean some knowledge. This cost will double if you can't find where you put those results. Simple job books, or folders organized in a recognized system and respected by all users is a critical key to success.

After departing from Boeing I did a small spin recovery program for Robertson Aircraft in Bellevue, Washington. That led to my analyzing and publishing performance data in the pilot's operating handbook for all of their Short Takeoff and Landing (STOL) airplanes. Then I did similar work with the Dee Howard Company of San Antonio for Lear Jets equipped with the Dee Howard Company thrust reverser. I hired a few Boeing engineers, either laid off, or those trying to broaden their experiences, and we began to do all of Lear Jet Series 20 airplane performance analysis. Larry (Mike) Timmons and Wade Roberts were key in that endeavor. They both had extensive experience in Boeing's aerodynamics group and that included performance analysis. With that experience they brought a new perspective to the Lear Jet, and the broader business jet market. We applied conservative techniques, or added safety factors, in analyzing the takeoff performance of an airplane with a failed engine. We published the results. We also incorporated additional conservatisms, or even greater safety factors, in the published landing distances. At that time this was a state-of-the-art endeavor. It was attractive for the manufacturers of these smaller airplanes to incorporate Boeing-style larger airplane performance analysis and data presentation techniques.

Our next contract was to produce the en-route speed-power flight operating (cruise planning) handbook data for the Allied Signal TFE-731 equipped Jet Star. This particular job was accomplished prior to the availability of large scale computer tools at SHANNON engineering, Inc. The work necessitated a great deal of manual curve fairing of the thrust and fuel flow required at various altitudes and ambient temperatures for specific cruise speeds. A serious problem arose in that the ambient temperature observed in the cockpit by the crew was slightly higher than the true ambient temperature because of the increase of that indicated temperature with Mach number. But our computer model of the engine, which predicted thrust and fuel flow, required true ambient temperature. Our solution was to draw massive curves representing the necessary data and interpolating for the solutions. The results were eagerly awaited as our predicted performance nearly doubled that of the original airplane: and, we were unknown in the industry. Almost immediately the crews reported that our predictions were consistently off! John Alberti, the brightest engineer I had ever worked with, noticed that they were only off on west bound flights! He wondered if the ambient air temperature probe was only located on the airplanes left side and consequently on west bound flights was exposed to the heating by the sun? It was, and suddenly we were heroes. We had credibility and we were the "go to guys" for airplane performance. Our Allied Signal work extended to performance analysis on all the H/S (BAe) 125 series and the Falcon 20 airplane with TFE 731 engines.

At about this time I hired our first CPA, Charlene Hinton. Accounting programs were expensive but she persuaded me to purchase and adopt an accrual accounting system. That was unheard of for a small company and a source of significant angst for our engineers. As we grew I added Jayne Weld Kaszycki to the accounting staff. Jayne's resume could easily have had only one entry: she passed her CPA exam the very first time!

In time employees wanted to buy stock in the company. I had originally incorporated in 1972 as Malcus, Inc., using my grandfather's name. We then incorporated as SHANNON engineering, Inc., so the employees could buy shares in the operating company while Malcus became the source of development funds.

By 1982, the time this story begins, we had a team in Sweden working on the international certification issues on the SAAB 340, and a consulting contract on the development of Canadair's Challenger. We had completed several helicopter performance improvement programs, and I also had a small contract with CASA, Spain's Aviation Company, previously known as Construcciones Aeronâuticas SA. We had completed the design and manufacture of camera bays in a variety of airplanes. And, we had a major contract with the Dee Howard Company to gain the structural certification of a modification to the King of Saudi Arabia's 747 SP.

Our strengths were that we could assemble a credible team to do almost anything in this field, and we had a good reputation, which extended to the international marketplace.

V The First Step

We were aware that the Boeing 707 and DC-8's were to be prohibited from operations by the congressional act of 1969 [A] and by the International Civil Aviation Authority, ICAO [B] in most nations beginning in 1985. The Port Authority of New York and New Jersey already had a landing noise ordinance as early as 1982. [C]

The work done in the 1970's by Boeing and others, including the availability of a large amount of public data, provided what seemed to be the basis for studies of a means of compliance.

So in 1981 we began a study of configuration modifications that might be reasonable to bring these older airplanes into compliance. Our company news reported [D] in October of that year that the study was underway.

One way of achieving compliance was simply to reduce the noise by reducing thrust. However, we suspected we couldn't lower thrust enough to still lift the necessary payload off the ground.

We could also make changes to the engine. But we believed that that would be very expensive and a great risk to the cost of development.

Another method would be to land at a reduced flap setting. A reduced flap setting requires less thrust to stay in the air and less thrust produces less noise.

Also, if you were to takeoff at slightly lighter weights you would be higher over the acoustic measuring station and produce less noise.

Finally, since we understood a significant part of the noise came from the front end of the engine, it ought to be possible to add acoustic lining (material to attenuate the noise) to the fan duct, and possibly to an extension of the inlet to sufficiently reduce the noise. Acoustic lining means material is added to the inside of the nacelle that actually attenuates the sound as it passes over the surface.

It was just a question of which combination of these modifications was considered the least risk, and the most cost effective configuration leading to successful compliance.

The October 1981 news release attracted quite a bit of interest in the operator group. I decided that the best thing we could do was to host a seminar and discuss the possibilities with operators. On February 23, 1982 [E] over thirty 707 and DC-8 operators showed up in Seattle to hear what we had to say.

They were desperate. I decided to be very honest and say that we could not complete any significant studies without funding, and if we were to accept any funding we could offer no guarantees. "No problem," they said. We collected $10,000 from five operators. I don't even remember any letters of agreement. That $50,000 went a long way towards our being able to define the market and the preliminary configuration.

VI Preliminary Configuration

We then embarked on a study of the most realistic means of compliance relating to the noise rules with John Alberti playing a principal role in configuring the nacelle modifications.

We concentrated on the 707 because we understood that market was larger [A] and that most of the attendees at our seminar operated 707's.

We used trade studies to show how combinations of weight reductions, nacelle treatments and reduced flap settings were required to reduce the noise for compliance and what the impact of these modifications would be on airplane pay-load and range.

We determined engine changes were not cost effective.

We completed a detailed study of the production airplane acoustic signatures from the public documents earlier mentioned. The phrase acoustic signature is generally understood to mean the noise level in one third octave bands in the spectrum of audible frequencies. Those signatures were measured at specific locations, or stations, around the runway for takeoff and landing. The offending noise came from exhaust gas velocities and from turbine noise emanating from the inlet and exhaust fan duct. Exhaust gas noise is caused by the mixing and tumbling of high velocity hot gas with ambient air. The turbine noise is caused by the number of compressor fan blades and velocity of the blade tips. The greater the number of blades and the faster the compressor turns, the higher the frequency.

Attenuation of the noise from exhaust gas velocity has historically been very difficult to achieve.

However, reducing the turbine noise was possible and there was considerable data available to consider this option. In this time period there were two materials that could attenuate turbine noise, DynaRohr made by Rohr Industries, Inc. and an identical design by Boeing. The materials are generally referred to as *acoustic lining*.

This material was comprised of two layers: a sintered metal tightly woven surface, or perforated plate, for an outer layer and a honeycomb internal structure. The woven outer layer was comprised of strands so small one had to look closely to see that it was not a solid surface. The depth of the honeycomb under the woven surface was tuned so that a particular frequency wave would enter the chamber exactly at the instant a previous wave was bouncing out. That is, the chamber would resonate at exactly the right frequency. Hence the strength of the wave adjacent to the surface was attenuated by the wave

VI Preliminary Configuration

We then embarked on a study of the most realistic means of compliance relating to the noise rules with John Alberti playing a principal role in configuring the nacelle modifications.

We concentrated on the 707 because we understood that market was larger [A] and that most of the attendees at our seminar operated 707's.

We used trade studies to show how combinations of weight reductions, nacelle treatments and reduced flap settings were required to reduce the noise for compliance and what the impact of these modifications would be on airplane pay-load and range.

We determined engine changes were not cost effective.

We completed a detailed study of the production airplane acoustic signatures from the public documents earlier mentioned. The phrase acoustic signature is generally understood to mean the noise level in one third octave bands in the spectrum of audible frequencies. Those signatures were measured at specific locations, or stations, around the runway for takeoff and landing. The offending noise came from exhaust gas velocities and from turbine noise emanating from the inlet and exhaust fan duct. Exhaust gas noise is caused by the mixing and tumbling of high velocity hot gas with ambient air. The turbine noise is caused by the number of compressor fan blades and velocity of the blade tips. The greater the number of blades and the faster the compressor turns, the higher the frequency.

Attenuation of the noise from exhaust gas velocity has historically been very difficult to achieve.

However, reducing the turbine noise was possible and there was considerable data available to consider this option. In this time period there were two materials that could attenuate turbine noise, DynaRohr made by Rohr Industries, Inc. and an identical design by Boeing. The materials are generally referred to as *acoustic lining*.

This material was comprised of two layers: a sintered metal tightly woven surface, or perforated plate, for an outer layer and a honeycomb internal structure. The woven outer layer was comprised of strands so small one had to look closely to see that it was not a solid surface. The depth of the honeycomb under the woven surface was tuned so that a particular frequency wave would enter the chamber exactly at the instant a previous wave was bouncing out. That is, the chamber would resonate at exactly the right frequency. Hence the strength of the wave adjacent to the surface was attenuated by the wave bouncing back out. Additional attenuation was achieved by the sintered woven surface through simple friction losses as the waves entered and exited. The classic name of such a chamber is a Helmholtz resonator. While the chamber under the surface was tuned to a particular frequency, the surface material acted to attenuate broader frequencies.

It was desirable to consider extensions to the inlet and fan duct to provide additional area for acoustic lining. If the acoustic lined inlet or the fan duct extension was too long it would add weight and aerodynamic loads to the nacelle, and too long an extension could compromise engine operating characteristics.

Engines operate on what is called an operating line. This line is actually a curve showing corrected air flow versus corrected fan speed. It is often best to think of that line as defining the angle of attack of the fan blades. If changes to either the fan duct or inlet should alter that line, the engine will be less efficient because the fan blades are not operating at their optimum angle of attack.

We made an educated guess. Our study was facilitated by a large amount of public literature allowing prediction of attenuation in acoustically treated ducts. [B] We had to consider the impact of inlet and fan duct extensions on flutter, loads, engine operating characteristics, and manufacturing cost.

Reverse thrust required the fan duct translating sleeve to move aft. That movement had been a problem on the basic 707. It moved reasonably well with reverse thrust but could become stuck in reverse and crews were known to carry a section of two-by-four lumber in the cockpit to hit the fan duct to get it to retract. Extending the fan duct to add acoustic lining could exacerbate that problem.

We had defined our preliminary configuration. We would incorporate all the acoustic lining that could be installed in reasonable extensions of the inlet and fan duct. We determined the amount of those extensions based on aerodynamic loads, flutter, engine operating characteristics, and manufacturing cost. We would modify the landing flap switch to allow landings with reduced flaps, and finally, we would accept take-off weight limits to gain compliance.

Our acoustically treated nacelle, the preliminary configuration, looked like this.

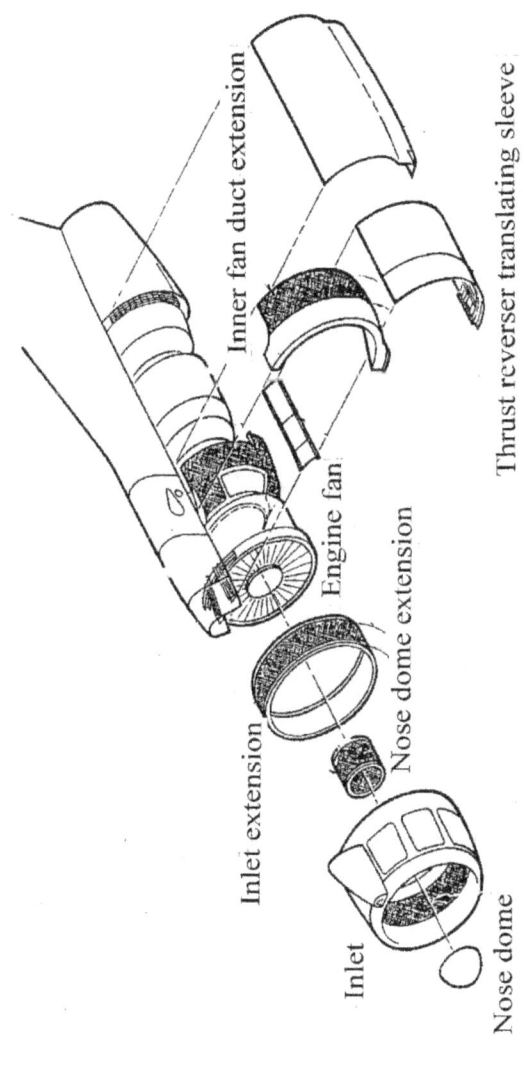

Quiet 707 Acoustically Treated Nacelle
Dark Areas are Acoustical Lining

Scott Toner spent hours crawling through a parked 707 at Boeing Field to determine how we might build a new reduced landing flap switch that replicated the features of the existing one, in addition to the existing functions required at a new reduced landing flap setting. It was necessary to retain the previously certified landing flap setting for emergency situations.

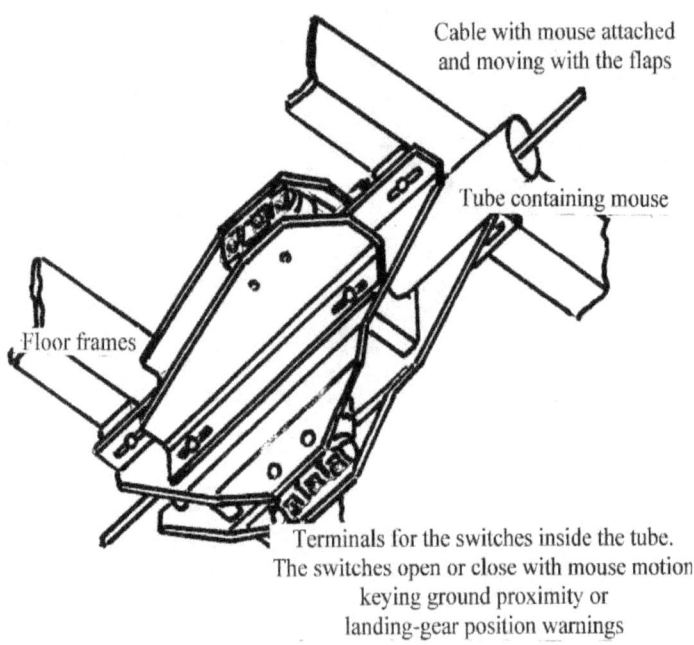

Quiet 707 Reduced Landing Flap Switch

The switch was a modification of the existing one under the floor of the cockpit and literally opened or closed, depending on the position of a mouse (or lump) on the cable which moved with the flaps. When the flaps were in the landing position the switch could alert the crew if the landing-gear was not down for landing and the switch would change the mode of operation of the ground proximity warning device.

Our study was released in a technical document "707 Noise Abatement Program, Phase I (Feasibility) Final Report."[C] The non-recurring cost would be about ten million dollars (actually it was 42 million dollars), and the best configuration was the acoustically treated modest fan and inlet extensions, a reduced landing flap configuration and a slight takeoff weight penalty.

However, there are variants of the 707. There is the 707-300B and C, the -100, a -138 built for Quantas and a short body, the 720. In addition there are several engine variations of the JT3D. Our engineering program would by necessity address all of these.

We knew we could demonstrate that 707's would comply with the 1983 Port Authority of New York and New Jersey noise ordinance if equipped with the reduced landing flap switch. We only needed to build a switch, install it in an airplane, and demonstrate the results to the FAA.

Manufacture of the switch required that we get a Parts Manufacturing Authority, or PMA, from the FAA. A Parts Manufacturing Authority assures all concerned that materials, processes and quality control are in place to fabricate an airworthy article, in this case, the switch. The PMA was granted July 14, 1983. [D]

Throughout this book airport identifiers [E] are used to identify the geographic locations of tests.

Performance and acoustic flight tests with the reduced landing flap switch installed, were proposed to the FAA [F] and were successfully accomplished on a Pan American 707-100B at John F. Kennedy Airport, KJFK, and at Roswell (Walker Air Force Base), KROW, in June of 1983.

The Supplemental Type Certificate for landing at reduced flaps, STC SA2063NM, was awarded on July 12, 1983. [G]

STC stands for Supplemental Type Certificate. A Type Certificate, TC, is awarded by the FAA to an airplane's manufacturer indicating that type, as defined by the drawings and processing specifications, is airworthy and may be used in commercial service. A Supplemental Type Certificate, or STC, is awarded for a product having met the same qualifications but is awarded to the group that modified the airplane. These certificates are the FAA's way of authorizing and controlling the production of aviation parts and assemblies.

We had successfully demonstrated compliance with the 1983 Port Authority of New York and New Jersey noise ordinance!

Naming our product proved to be a challenge. In a Dallas bar with Bob Wildby, Gates Lear Jet, and Mike Conlin, consultant, and after a round of drinks, I came up with the name "Quiet 707" or "Q 707." It stuck.

We then embarked on an effort to raise the ten million dollars that we thought would be required to modify the nacelle. We had just entered the area of our greatest ignorance.

VII The Market for Venture Capital

The 1980's heralded a unique period concerning the tax treatment of corporate spending for research and development. The Economic Recovery Tax Act of 1981[A] basically allowed an accelerated tax credit for contractual commitments to fund specific research and development tasks. The act was intended to stimulate business, which it did. By 1990 the tax code was rewritten and the consequences of the Act have been studied extensively. [B]

What the act did accomplish was to stimulate funding for developing programs, which the Quiet 707 certainly was.

However in the same decade, aviation was subjected to other threats. One writer noted "During the 1980s tort lawsuits decimated the general aviation industry." [C] By the early 1980's employment, sales and manufacturing rates were seriously impeded by these personal injury law suits. This same writer noted that employment dropped by 65%. It has been estimated that one dollar in three in the purchase price was due to litigation expenses. That was a serious impediment to the development of new aviation products. [D] As a result aviation companies divested into as many other venues as possible in order to hedge their risk. This resulted in companies making acquisitions of other companies that seemed a good diversification only to learn there was often a good reason that the company could be easily purchased. The buyer could do no better at managing the company and often the acquisition became a financial liability.

Basically these two factors created an environment of empty hangars and production facilities ripe for a credible research and development program. This was just the environment we needed to launch the Quiet 707 program!

VIII Seeking Research and Development Capital

There were two unique characteristics of the Quiet 707 program. The market was readily identified, and we had a close working relationship with our customers through correspondence and seminars. [A] However, this group could not be a source of development capital. Our estimates in JS1106, [B] which proved to be completely naive, was that the program would require ten million dollars in non-recurring capital. Non-recurring capital is the amount of money to develop the design, produce tooling and be in a position to produce the first certified article.

Venture capital typically is used to develop products with much broader markets. It sounds easy at first. You "arrange" for someone with the right connections to raise the necessary capital. You read about that all the time in the newspaper. In fact, there seems to be no shortage of money lying around just waiting for the right opportunity.

The first problem is that the venture capitalist wants an exclusive on your program. That means among other things that you should not be out trying to raise capital on your own. In fact you probably cannot negotiate a funding contract because the venture capitalist believes he has every right to negotiate that contract and he will get a part of the deal. The next problem arises because you really have no way to ascertain that the capitalist has the right connections or even the necessary ones. The next challenge is that with an exclusive contract the venture capitalist can sit on his hands and sign agreements with other venture capitalists believing that the more people on the job the better the circumstance that the money will be raised. Now, the final problem is that none of these people really understand your market or the program. They will unknowingly misrepresent the potential gains and your ability to achieve those with minimum risk.

There is one other requirement. The venture capitalist is happiest if you have mortgaged your home, spent all of your life savings, sold your daughter into slavery and desperately need the program to get back on your feet. That way he has every assurance that you will work very hard to be successful, and he, in turn, can make money off of your efforts.

One venture capitalist told me he expected a return in five years of fifteen to one on his investment, and if we failed he would expect the revenue from our next three programs.

So why would you use venture capitalist? If your market is narrow, by that I mean it addresses a niche with a very specific product, I would not do so. You know that market best, and you understand the risk. The only problem is that you may not have the knowledge in structuring a financial deal. Here, in this instance the funding group and the market only want to know one thing, "who do we go against when things go wrong?" You didn't have any money in the first place, so it's not you.

What you will find is the eventual funding organization will have a need, and what you propose to do can satisfy that need. There can be harmony, mutual needs, or matched impedance, in being involved with them. The relationship starts with a feeling of mutual trust. You will need to understand their need in great detail.

Just look at what the stock market is delivering as your competition. The investor can get those earnings without the risk associated with your program. Your program must deliver much more earnings on their investment than the market can.

We made a list of those companies that had the expertise, capacity and possible interest in funding the Quiet 707 program. E System in Greenville, Texas; Gates Lear Jet, GLC, in Tucson, Arizona; Tracor Aviation in Goleta, and Rohr Industries in Chula Vista both in California were the candidates: E Systems because of their prominence in after-market major structural changes to large airframes, usually for domestic and foreign governments; Gates Lear Jet because the litigation industry had all but shut down production and they needed a hedge business; Tracor Aviation because Tracor, Inc had purchased Tracor Aviation and found it too difficult to secure the predicted business; and Rohr because they had, or should have had, all the tooling for nacelles for the Boeing 707.

Our hopes of entertaining E Systems never amounted to anything. They saw their usual business as having much less risk than the Quiet 707 program.

Gates Lear Jet, GLC, had suffered at the hands of personal liability lawyers. Their production was down and sales were off. They had a factory which was not being utilized to the fullest in Tucson Arizona. Through Mike Conlin, I was introduced to Jerry Gilmore an officer at GLC. He offered $25,000 for an exclusive on the Quite 707 program. That would allow time for GLC to conduct their due diligence. If that study proved that our program was a good match for GLC they would then enter into a contract to produce the kit. At the end of that period in a meeting with Bib Stillwell, GLC President, there seemed to be some confusion as to exactly what GLC had agreed to and what the expectations were of me; even though the agreement struck by Gilmore and myself was a written one. This was very detrimental to us as I had devoted so much time to the program that our other sources of revenue had suffered and I was at risk of losing key employees.

I distinctly remember how tired I got of going down to Tucson to be disappointed, that I secretly purchased flowers and asked a secretary to decorate the conference room in the hope I could bring some levity to this exhausting situation. Immediately on entering the room one of the GLC officers removed the flowers and asked "Who died?"

But GLC's tentative nature should have served as a warning. I had been as busy as a person could be just running SHANNON engineering, Inc., and that time taken had a serious impact. I didn't have a chest of money to make payroll while I responded to every potential funder's whim. We were not only losing business, we would soon lose employees. Replacing those people would take critical time. The longer a capitalist or funder can string you along the more desperate you become, and the more favorable the deal will be to them.

Rohr's interest was expressed by a visit with their representative, who on looking out the window of my office to the Boeing Field flight line, asked "Where is your factory?"

Our basic but serious problem was that we hadn't found the company with a need that could be satisfied by our program.

IX Contract Negotiations

Tracor Aviation began to take a hard look at our program. They had recently completed the Super Guppy program and the Quiet 707 project seemed no more challenging.

I began to supply schedules, lists of clients, estimated program costs, and risk factors to Kirk Irwin, Vice President, Engineering and Development for Tracor Aviation, Santa Barbara, CA. As he developed an interest with Tracor, Inc., I traveled to Austin, Texas for many visits with the parent Company.

I couldn't have been happier to have Kirk's attention. By then I was a little thread-bare. Kirk had been described by a friend at the FAA as someone who, when everything else had been a complete disaster, can stand up, brush himself off, and say with confidence "now here is what we have to do." I had reached the point where I was beginning to feel as if the Quiet 707 program had been a complete failure owing to lack of funds.

Unknown to me at the time that I was making these visits, Tracor, Inc was having similar discussions with the competition, an A. B. Stuart of Waco, Texas. Mr. Stuart was a retired executive of American Airlines and now offered the 707 market exactly the configuration described in the 1970's work by Boeing, McDonald Douglas and the FAA. To claim that configuration as viable was just not accurate.

Tracor understandably had timidity in embracing the Quiet 707 program. We were a relatively unknown source. William A. Conner, Tracor's resident acoustician, was tasked with a study to determine the feasibility of our program. The study incorporated one hundred and eighteen technical acoustic publications as references. Mr. Conner concluded "The Shannon report does not demonstrate the feasibility of achieving the projected acoustical attenuation using the proposed nacelle treatment." [A] It was the only conclusion he could reach. Had he said the attempt would be successful and that had not been the case, he would certainly have lost his job. On the other hand, if his conclusions proved to be wrong it would only prove to be embarrassing for Mr. Conner.

Something else was going on in this time period. Tracor, Inc. was our last hope. I spent so much time courting them that I began to neglect SHANNON engineering, Inc.'s business as usual. Revenue fell and the employees began to sense we weren't going to be successful. As Tracor continued with their due diligence I lost a very special employee, John Alberti. He was the most creative and imaginative acoustical engineer I had had the privilege of working with. I couldn't blame him. It was the smart thing to do. But the very serious casualty was the ability of SHANNON engineering, Inc. to survive if we didn't get funding. I knew how difficult it was going to be to hire new individuals and get them working as a team.

X Contract Signing

Tracor had a fundamental reluctance to contractually commit to finance a program that would perhaps reduce their cash drain at the Santa Barbara facility. They were introduced to Morris and Douglas Jaffe through a venture capitalist. The Jaffe's said that they had a ten million dollar investment opportunity that had "stalled" with the owner of the New Orleans Saints. That ten million dollars was available for a deserving and attractive program. Since Morris and Douglas claimed that they had had a previous involvement in developing aviation programs, the Quiet707 program seemed a good fit and they were willing to put up the non-recurring capital. If their capacity and interest was as represented, it was a deal made in heaven for Tracor. Since these discussions were in confidence with Tracor I never knew if Tracor did any due diligence to confirm the Jaffe's credibility. It is my guess they did not, as the Jaffe's offered a no risk all reward opportunity to Tracor.

From the beginning I had been informed by Tracor that this would be a very expensive program. I learned that Tracor had approached at least one competitor to see if a more favorable deal could be struck despite our close relationship with the market.

Finally after too many trips to Austin, Tracor and I were ready to sign. I would get paid for all our work to date, a sum of $150,799 which included reimbursement for the 707 inlet we had purchased; SHANNON engineering, Inc. would be recognized as having conceived of and of initiating the program; we would have an exclusive on all engineering; we would be the FAA Supplemental Type Certificate, STC, holder; and I would receive royalties on each delivery of a Quiet 707. The contract was signed October 25, 1983. [A]

Signing with Tracor, Inc. provided enormous relief from what I considered to be huge administrative responsibilities. Tracor Aviation would be the installation center, the point of sales, provide overall program management and provide product support to the customer. I knew SHANNON engineering, Inc. would be busy just with the engineering work certifying the configuration. Tracor Aviation also had, in the person of Kirk Irwin, plenty of experience in airplane modification programs, something Tracor, Inc. did not have. So Tracor Aviation became an insulating group as SHANNON engineering, Inc. prepared for what would become its largest ever program. Such programs are fraught with frustrations, poor weather, delayed schedules and even failures. We had the best of all worlds by not having to answer to Tracor, Inc. for every unexpected event.

There seemed to be an opportunity to ease the cash flow situation for both Tracor and Community Transports, Inc. if I split the royalties between the two. Since Community Transports, Inc. had convinced us they had developing interest in aviation programs, I could acquire a perpetual exclusive on all that engineering work in lieu of half of the cash royalty. I signed that agreement on November 28, 1983. [B]

I could go home to Seattle and start hiring engineers to complete the program.

XI Development

The program outlined in our feasibility report [A] required two-and-a half years to deliver the first ship set, or single airplane kit. It was necessary to start the program in June of 1982 to meet the deadline on January of 1985. That early period would be devoted to structural analysis, prototype nacelle fabrication, engine and acoustic ground testing and negotiation with the Federal Aviation Agency (FAA) as to how much testing and documentation would be required. That would take us up to August of 1983. From that time to January of 1985 we would be in flight tests and certification documentation which would be completed according to the previously agreed plan. To have finally obtained funding as late as October of 1983 reduced our program to fourteen months rather than our planned two-and-a half years. This definitely presented a challenge to us.

Signing with Tracor, Inc. had enormous benefits. They were successful in contracting with Rohr Industries, Inc. to complete the detail design and build the hush kit nacelle components. It turned out Rohr did have the tooling for the original nacelle, and the fact that they were responsible for that design gave enormous credibility to the Quiet 707 program. Our design was nothing more than a layout, or sketch, of the necessary parts.

A nacelle is perhaps the most complex part of an airplane: it is subject to vibration, and high temperatures, it must be flame retardant, it is subject to aerodynamic loads and it must be durable; but then it must easily come apart for maintenance. The engineers at Rohr not only had plenty of experience in that field, they had that with the 707 nacelle! We could not have done any better than having Rohr as a partner. They gave the program widely recognized credibility in a difficult market place.

Tracor decided they would charge three million dollars for the kit. Of that they would get about one million and one hundred thousand dollars as the installation center, for program management and they would pay for our engineering services. They would also stand behind the product for warranty and liability claims, be the lead sales organization, and coordinate with Rohr as the nacelle manufacturer.

As the applicant for the Supplemental Type Certificate, STC, I licensed Rohr Industries to build the nacelle parts. It had previously been determined that they had sufficient tooling, called rotary tooling, to build multiple nacelles simultaneously. This aspect was a key item because we needed to capture as much of the market as we could in a short time period. We would be required to build four nacelles to deliver a single airplane hush kit.

This together with the reduced time to penetrate the market dictated a requirement for rapid production of multiple kits.

SHANNON engineering, Inc., went from about seven employees to twenty and we were spread too thin. We were under contract to the Swedish company Svenska Aeroplan AB, SAAB, which required two people to spend their time in Sweden. We had other employees in San Antonio under contract to the Dee Howard Company completing the design and analysis of modifications of the King of Saudi Arabia's 747 SP. Obviously, competent experienced engineers are infrequently available and never when there is a need. We were severely understaffed. I did the best I could, and it worked. That is really the final test.

We had an agreement on the amount of test and analysis required by the FAA for certification of the Quiet 707. In all, some one hundred and fifty technical reports, many with supporting appendages, would be prepared and submitted to the FAA. The reports addressed structures, systems, acoustics, airplane and engine performance, and flight manual presentations. A vital report was the compliance document [B] which detailed where each compliance with every regulatory requirement would be found.

From October of 1983 to the following March Rohr Industries, Inc. was busy completing detailed design, fabricating tooling unique to the hush kit, and building the components of first Quiet 707 nacelle.

By April of 1984 we were ready to have the prototype nacelle hardware installed for ground tests at Rohr's test stand at Brown Field, San Diego. [C]

The Quiet 707 Engine and Nacelle on the Test Stand at Brown Field, California.

These tests were comprised of acoustic and engine performance ground tests at various power settings.

Acoustic signatures are recorded by arrays of microphones around the test stand. It is then possible through the use of computer programs to add the effects of forward airplane velocity and altitude to those recorded acoustic signatures. It is then possible to replicate with great confidence the noise expected when conducting the specific FAR 36 demonstration flight tests.

The purpose of engine performance tests is to assure that the operations of the engine have not been affected by the additional inlet and fan duct length. Test results confirmed the inlet and fan duct extensions had not altered the engine operating line!

With the successful completion of the ground test, we knew we were home. Together with the previously approved ground proximity switch, this meant we were ready to go into production. The acoustic signatures from Brown Field could be translated into flight to later be confirmed by flight tests. Now we knew with some certainty how much weight we needed to off load to meet compliance with the Stage 2 requirements of FAR 36.

XII Flight Test

We had selected Moses Lake Airport, Washington (KMWH) as our preferred acoustic test site. Putting the acoustic flight test first revealed Tracor's uncertainty about our ability to comply with the requirements.

There were two alternatives: Goodyear (KGYR) with the airplane and crew based in Phoenix, Arizona (KPHX), and Albuquerque, New Mexico (KABQ). Again, I have used airport identifiers [A] to identify the geographic locations of tests.

We had made inspections at each of these alternatives and believed that if Moses Lake did not work out for any reason, one of the alternates would be suitable. We had communicated with Tracor that the end of October was the latest we could find acceptable weather at Moses Lake.

Selecting a site for flight tests requires meeting very particular requirements: the weather must be clear with no clouds or gusty winds, a runway as long as ten thousand feet is desirable; broad unencumbered areas should be available adjacent to the runway with plenty of flat ground all around; and finally, because of the number of people involved, a city with adequate motels, restaurants and rental cars must be near. Those requirements are most often met by an abandoned military field. The acoustic requirements included narrow humidity and temperature lapse rate (the change in temperature with increasing altitude) which further narrowed the possible tests sites.

Typically we flew someone else's airplane under contract. This always required some familiarization with our pilot in the left seat and the owner's pilot in the right. Our test crew was Jim Gannett and Jim Goodell as pilot and co-pilot. They were more senior than most, and out of hearing they were callously referred to as the geriatric crew. However, after an introductory flight, the owner's pilot returned to the ready room, shook his head, rolled his eyes, and said "these guys can really fly the airplane!'

Jim Gannett was the nicest guy you could ever hope to fly with. He had been Boeing's Supersonic Transport Project Pilot, and he was co-pilot when Tex Johnson rolled the 707 over Seattle's hydroplane race course in 1955 (a very unpopular act with Boeing management but a great publicity stunt).

Jim had a dry sense of humor. I remember one occasion when Boeing was testing navigation features on the AWACS (the 707 with the very large radar dome on the top). An important test of the AWACS was to quantify the accuracy of the navigational system. The intelligence obtained from the radar was only valuable if one knew where that intelligence had been gathered. That task is more complicated if it is not exactly over or exactly between fixed navigational points on the ground. In this test those points were VOR's, or Visual Omni Ranges, which transmitted navigational radio signals from known locations to the airplane. Jim was asked if he could "just miss" going directly over one of the fixed navigation points on the ground. Jim replied "I don't even have to go there!"

Jim was everything you could want in a test pilot. Easy to work with, a good communicator, and a very strong team player. He was not the type of pilot who would go off and do something not in the plan, something he felt to be important, and in so doing expose the airplane and crew to unnecessary risk leaving the engineers to figure out what to do with the results.

Before any flight tests could be accomplished the structural analysis of the entire airplane had to be completed. The analysis addressed the additional weight of the nacelles including the acoustic lining, and the aerodynamic effects of the extended inlet on airplane loads and structural integrity. When that analysis was provisionally accepted by the FAA we could begin our flight testing

Galen Haws, our structural engineer, came up with a novel and very useful manual note template. It is difficult to understand for the uninitiated, but the volume of testing could easily lead to confusion as to exactly what was intended at a particular moment in a given flight. Manual notes have historically been used to record what tests were being accomplished and in what order. Manual notes were also valuable in recording anomalies or system failures that might affect the data obtained in the test. So simple a tool, and so terribly valuable!

The first flight tests would be to demonstrate that the airplane was free from flutter. The test was comprised of various excitations (combinations of rudder, elevator and aileron "kicks" or pulses) at selected speeds and altitudes with combinations of fuel loads. Flutter tests are critical flights, and are only accomplished with a minimum crew.

Flutter tests were successfully conducted on our test airplane, N886PA, a 707-321B, at Mojave airport Lancaster, California (KMHV) October 24, 1984.[B]

Our flight engineer for the flutter flights was Wayne Miller. Wayne's nickname became "Thumper." His voice was the first we heard on the radio after each critical high speed flutter dive, and he reminded us of the rabbit character in Snow White. We were beginning to feel very good about our program.

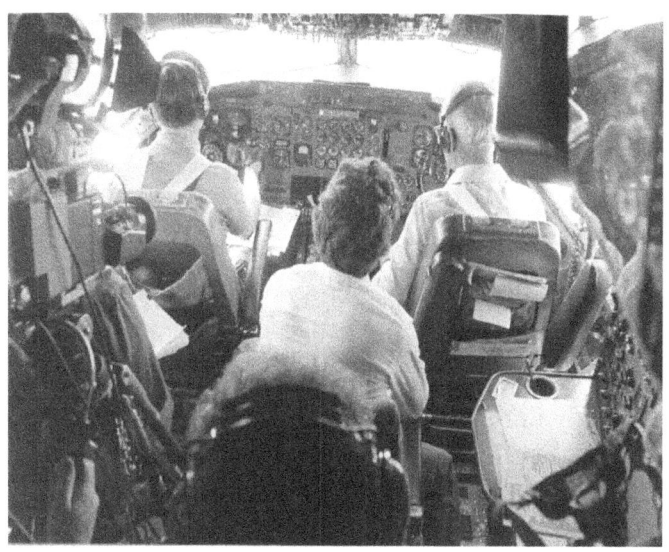

Test crew in the cockpit.
Pilot Jim Gannett, Co-pilot Jim Goodell,
and Flight Engineer Wayne Miller.

However, the availability of the airplane for acoustic tests was delayed. We prepared for tests at Moses Lake in early November, just outside the calendar envelope we required. The very day we intended to begin tests a weather system set in bringing freezing temperatures, ice, and gusty winds. We were shut down and it wasn't going to get any better.

I immediately set out for Goodyear to make arrangements to test there. Relocating the acoustic test site required moving a large amount of ground based equipment and as many as thirty personnel. We had only days to get an agreement for operations and flight tests with the Goodyear airport manager, rent vehicles including a mobile home for an onsite office, make arrangements for lodging, airplane fuel, survey acceptable locations of the microphones, obtain a power source for those installations, rent a light airplane and install instruments for airborne atmosphere samples, contract with a pilot for that airplane, and install the meteorological tower required for acoustic tests. I also took doughnuts down to the local shopping center just off the end of the runway to acquaint residents that tests were planned that would produce significant noise from a low flying Boeing 707.

Seven days after abandoning Moses Lake we were installed at Goodyear and ready to begin our series of acoustic tests. Those tests were successfully completed between November 13, and the 15th 1984.

Test aircraft in the early morning over measurement station at Goodyear (KGYR), Arizona. Galen Haw's is in the light shirt.

The remaining tests were seen as low risk. We would demonstrate stall speeds, or the lowest speed at which the airplane could fly. Takeoff and landing speeds are based on safe margins above those speeds. We would demonstrate good engine operating characteristics, and that neither the inlet or fan-duct extension had adversely affected engine operations. These tests would include rapid throttle movement at every adverse flight condition imaginable. We would even include tests at the world's highest airport, La Paz, Bolivia. Tests would include demonstrations of autopilot operation, lateral-directional control, cross wind landing demonstrations, takeoffs with incorrect trim, and landing light evaluations. Maneuver margins were evaluated (a demonstration of how hard you can turn the airplane without encountering stall), go around engine performance (a simulated aborted landing), and engine starts in the air. Some 21 separate flight test analysis documents were planned to address FAR compliance matters.

These tests were accomplished at Santé Fe (KSAF) and Roswell (KROW) in New Mexico, and at Santa Barbara, California (KSBA), in addition to those at La Paz (LBP).

Flight Tests were completed on the basic 707-321B by December 5, 1984.

XIII Certification

The necessary testing and analysis to document the airworthiness and compliance with acoustic standards would require over one hundred and fifty documents, many with several volumes of appendices. The organization of that library was tracked by a "drawing tree" measuring 60 inches by 36 inches showing the documents grouped by compliance methods: acoustics, performance, structures, flight test, and Flight Manual Supplements. The compliance document [A] gave the overview as to whether compliance was to be shown by test and analysis, just analysis, or by "identicallity" to the previously certificated 707. Identicallity is a colloquialism meaning the modification produces no change from the original basis of compliance. The 707 having made its first flight in 1954 was certified under Federal Regulations CAR 4b dating from 1953. Accordingly, our compliance document referenced that set of regulations and any other (current) Federal Aviation Regulations the FAA deemed necessary.

Somewhere along the line we began to get very peculiar questions from the FAA. Our relationship was so close, and so well coordinated that we thought we knew every question before it was asked. But these were strange. They were from outside the normal certification envelope. It was if whomever was asking, was not that familiar with the program or certification methodology. Finally I asked the FAA individual most responsible for our program where these questions were coming from. To my surprise he answered that a Washington D. C. lobbyist, Hector Alcalde, had been hired to "monitor" our progress. He had obtained enough knowledge of our program to ask some basic questions. When I asked my friend who would hire such an individual his answer was that he most likely had been hired by competitors in the business to impede our program.

Acoustic compliance with domestic Federal Aviation Regulation Part 36 and foreign ICAO standards was achieved by January of 1985.[B]

STC SA2699NM [C] was issued March 6 1985 for the Quiet 707-300B with JT3D-3B engines.

Certification of subsequent configurations was accomplished by issuing supplements to the Flight Manual. These covered operations at La Paz (the world's highest airport at 13,325 feet), weight limits for the unique conditions of Nairobi (unusual combinations of high altitude and temperatures), JT3D-3C and 7 engines, intermix engines (not the same type of JT3D in any of the four locations), flight with the spare engine pod, three engine ferry, reduced takeoff thrust (lower thrust provided longer engine life), over speed takeoffs (faster

at lift off speeds allow higher takeoff weights) ,
compliance with the ICAO Annex 16 Noise Rules, and
for operations at non-noise limited airports.

The 707-100B STC was issued September 26, 1986 [D]
and the 720B STC was issued the following September
11th, 1987.[E]

In total, we had accumulated eighty one hours of flight
tests and thirty-one-thousand five-hundred man hours of
engineering and ground support time.

XIV Business Dealings with Boeing

We frequently had strained relations with Boeing. I had worked at Boeing, as had a number of our employees. It was impossible to keep that relationship at arm's length. We tried to keep a professional and responsible attitude because we never knew where we might need assistance. Some at Boeing thought we were providing Boeing a service, in that continued operations of the 707 protected the market of the older airplanes, which would open up the market for newer Boeing airplanes. Others were of the opinion, and I believe they were serious, that our interest was only to obtain Boeing proprietary data. This group also believed their best penetration of the market for new airplanes was to buy these older 707's and cut them up for scrap!

One 707 operator actually threatened to sue Boeing if they did not fully cooperate with us! That operator had no formal relationship with us at all and his demand seriously complicated any communication we would have in the future. These inquiries, either those formally from us, or from well-wishers not connected with us, created considerable difficulties as shown in the following letter.

ELAINE,
PLEASE RELEASE THIS TO PROJECT & STAFF SURVEILLANCE DISTRIBUTION.

Thanks
D. McGlothlen
8/7/5

August 5, 1985
6-1171-RKD-1237

To: C. A. Smith   6C-96
         D. C. McGlothlen   69-93
         R. R. Downing   6A-03

cc: C. R. Bradenburg   02-62

Subject: Boeing Proprietary Data

In the past three or four weeks Contracts and Software Sales has had requests from five different sources for proprietary data related to the 707-Quiet Nacelle program. These requests ranged from engineering firms, customer (Ports of Call) and a U. S. congressman.

We have reason to believe that some of these sources may use their knowledge of Boeing employees or use some other ruse to obtain such data.

Particular requests which we recall, other than those straightforward written requests, are from Far West Special Products and Shannon Engineering.

Please alert your personnel to this tactic.

R. K. Dings

One Example of Warning Communications
Within the Boeing Concerning
the Quiet 707 program

XV Production

"the Position Report," the company's news letter, reported in January of 1986 that we had firm orders for 79 deliveries at 5 per month. [A]

But the Federal Noise rules were not endorsed by all. Some communities exposed to the noise of night time cargo deliveries and departures drafted their own noise rules. Operations of the Quiet 707 were prohibited by the San Francisco Airports Commission in the fall of 1985. This was so threatening that cargo carriers funded a study concluding "they (the local governments imposing unique ordinances) interfere with, or prevent, the free movement of interstate commerce." [B] We were astounded by this, having invested so much to comply with the Federal standards! In fact, the Federal Government did little to discourage local ordinances. The San Francisco Airports Commission ordinance was not overturned until 1988. [C]

Then Congress, responding to those operators who had not taken delivery of their installed hush kit, passed an exemption or waiver recognizing intended compliance with the noise rule "if the operator had a valid contract for delivery of same." There then bloomed at least twelve companies offering "valid contracts" for basically the configuration described by Boeing from 1973 and

1974. One was so bold as to use our art work in their promotional brochure. Our attorneys wrote to them suggesting they not do that. They then offered to sell us their program. And, they claimed to have a large number of customers! Neither their program nor their claim of customers was credible. But that activity did create confusion in the market.

The work done by Boeing and others in the 1970's and in particular the DER form signed by a Boeing engineer stating that the design resulting from those studies was "capable of being certified" caused the most trouble. Several companies were proposing to build that exact package. One company in particular was headed by a retired airline executive who was well known in the industry. His efforts appeared to be credible. It was very difficult to inform prospective clients of the difficulties in pursuing a package that was "capable of being certified." Some groups proposing to build a kit had no concept of that task beyond offering a "valid contract" for a fee.

All these acts caused confusion in the market place and effectively slowed our delivery schedules. The 707 was ever getting closer to the end of its useful life. The most cost effective way to complete the program was to deliver as many kits as we could in the shortest time period. Congress and these opportunistic groups worked against that, and seemed to sabotage our efforts.

The first delivery was to Independent Air, (also identified as KAL Air) registration N728Q, S/N 20025, a 707-321B, on April 10, 1985. The last was to Buffalo, registration N1108BV, S/N19177 a 324C on January 2, 1990. [D] One hundred and seventy two hush kitted aircraft were delivered through 1991. Many were of foreign registration.

By April of 1989 "The Boeing News" credited us with having increased the total number of 707's in worldwide operations. That was quite an acknowledgement for our small company.

An unexpected benefit and a valuable education resulted from my time spent with the Quiet 707 salesmen, Pat DeYoung and Bob Watson.

As an engineer I had understood, or at least thought I did, that any product must be the best compromise of all the physical and engineering challenges. And having achieved that, the product must sell because it did just that.

But that is a sad misunderstanding. Previously I had had a contract with Mitsubishi. At that time their product was the MU-2, a very high performance turboprop airplane. The high performance was achieved in part by a short body and small wing. These two characteristics provided lower drag compared to other airplanes. That was achieved by reducing the surface or wetted area which produces friction, or drag. The down side was the smaller wing and short body produced a short coupled, or "goosy," airplane. In other words it had some characteristics that piloting it could require constant attention. Mitsubishi hired an individual to sell the airplane in South America. At that time that market was comprised of short rough fields with difficult approaches. That was not a comfortable environment for the MU-2. If you imagine an airplane with very good, docile slow flight characteristics with big tires; and the propeller or engine well off the ground and not subject to picking up rocks; you have a good airplane for the South American market. That was not the MU-2. The man responsible for those sales told Mitsubishi "Will you build an airplane I can sell, or do you expect me to sell what you are building? There is an enormous difference!" My point being the airplane was a wonderful example of balancing the requirements of aerodynamics and physics, but the design engineers completely lost sight of the peculiar needs of a broader market.

The Quiet 707 had its own unique market needs. There were numerous competitors offering to build hush kits. All were in the contracting phase with the "design" quantified by brochures of promises and grand assurances. The US Congress would easily offer an exemption from the requirements of the Federal Aviation Noise Abatement Policy providing there was a valid contract for any hush kit. So exactly why would anyone buy the Quiet 707 package? Grasping that fact identified the salesman.

Pat DeYoung could expound at length about the individual operators. What their past experience had been, whom they considered their associates, and what aspects of their operation should be considered sensitive. Not all the operators were charitable, ethical individuals. Guns, drugs, even slavery seemed to be in the wings of the theater. But Pat took all this in his stride. He could point out why the Quiet 707 was the only investment that assured the value of the 707, its continued operation, and the underlining value of the operators business.

Bob Watson was nothing short of a piece of work to watch. We made two trips to China in the late 1980's. We were visiting the Chinese Aviation Administration, or CAAC, a government entity. CAAC had ten 707's. We met in the old opera house in Beijing, an old stone building with pleasant meeting rooms furnished with free standing lamps with shades, overstuffed chairs with faded embroidered doilies on the arm and head rest. We would meet in the morning for social conversations, then have a very large buffet lunch served on an equally large lazy susan over a round table, then we would meet for business. Suffering from jet lag and the large wonderful lunch, we were easy prey for the Chinese! Bob noted that after lunch on the second day. There was an outburst of boisterous robust laughter! Don't miss the point: if you can tease your host and client you have established a very good relationship. You understand that market. Bob made the sale.

There is an art to selling. And it is easily as important as meeting the engineering challenges.

XVI International Aircraft

The world map of section I, page 5, which was used to stimulate interest in funding, shows the dramatic worldwide change in tolerance for airplane noise in just three years between 1985 and 1988. Many more countries were expected to impose noise standards by 1988. The map does not give a complete picture as the 707, having a range of about 6,000 nautical miles, could hardly be expected to operate profitably only in those countries without a noise ordinance. Mike Conlin's study [A] of 1982 showed that of the 507 candidate airplanes there were 377, or almost three quarters of the fleet, that were registered in 110 foreign countries. The good news was the noise criteria was the same, ICAO Annex 16. [B] But, each foreign government could be expected to complete its own airworthiness investigation including flight tests of the Quiet 707. Further complicating the market penetration, within that group of foreign operators there were 86 operators with fewer than five airplanes, many of those airplanes may not have constituted a suitable airframe for the Quiet 707 hush kit. To be suitable, an airplane had to conform to the configuration that Boeing delivered and it had to have an acceptable maintenance record.

We obviously were going to have to cherry pick the candidate airplanes and would expect to spend considerable time in certifying to foreign standards.

Somewhere in this time period I met operators whom I was sure were moving guns to troubled areas and returning with illegal drugs. I heard of, but did not see, a slaver; an airplane with cages for moving people; slaves. One customer had an airplane we called "tuna fish." He was the chairman of the board of a significant Houston based international oil and gas company. He wanted to sell his company's 707 to a particular African head of state. To do this he furnished the airplane with booze and prostitutes dressed as flight attendants. It smelled like fish.

We demonstrated compliance with the noise rules of ICAO Annex 16 in January of 1985.[C] This was the first time that a domestic noise STC had been accepted by a foreign government. [D] Airworthiness matters were first addressed with The United Kingdom's approval in 1986.[E]

In time we would deliver eighty-eight Quiet 707 aircraft to twenty foreign operators. We demonstrated compliance with foreign agencies in Angola, Belgium, Chile, Columbia, Ecuador, Egypt, England, France, Iraq, Israel, Pakistan, Peoples Republic of China, Portugal,

Romania, United Arab Emirates, Saudi Arabia, Sudan, Switzerland, Uganda, and Zimbabwe.

Now thirty years after the inception of the program it is interesting to look at some of the clients in the light of today's international relations.

民航處
航空安全事務科
適航標準及機航事務組
辦事處：香港國際機場停機坪大厦 259 室

CIVIL AVIATION DEPARTMENT
AIRWORTHINESS & OPERATIONS SECTION
AVIATION SAFETY DIVISION
Room 259 APRON SERVICES COMPLEX
HONG KONG INTERNATIONAL AIRPORT
HONG KONG

OUR REF: AD/500
YOUR REF:
TEL. NO. 3-7697641
CABLES: AVSTANDARD HONGKONG
TELEX : 39524 CFSHK HX
AFTN : VHHHYACA

22 January 1988

Mr. C.S. Fender
Shannon Engineering Ltd
7675 Perimeter Road South
Suite 200
Seattle
Washington 98108
U.S.A.

Dear Sir,

Boeing 707-336 (Con. No. 20517) Reg VR-HKK

Thank you for your letter of 21st October 1987 regarding the Shannon/Tracor modification to the above aircraft.

I agree with your statements regarding the similarity of the B707-330C & 336C. Hence Hong CAD Airworthiness Approval Note No. 420 is equally applicable to Con. No. 20517 (VR-HKK). The AAN itself in para. 10.0 embraces all the -320C series and hence needs no amendments.

I also therefore agree that the only action required is for a JS 1263 AFM Supplement to replace the JS 1185 currently in the aircraft manual.

Would you please make arrangements to supply one copy for the aircraft manual and one for our office copy. This office will take responsibility for effecting the change to the manual.

Finally my apologies for taking so long to reply to your letter - we have actually only just issued the Hong Kong C of A to the Aircraft which is now in operation with AHK Air Hong Kong Ltd.

Kind regards to all at Shannon.

Yours faithfully,

(D.G. Haward)
for Director of Civil Aviation

c.c.: Mr. H. Goldberg - Tracor Aviation
      Mr. I. Robertson - AHK Air Hong Kong

CCSA

Pakistan International
Head Office :
PIA Building, Karachi Airport
Telephones : 412011 (Admn.)
439259 (Engg.)
Cables : Pakinfair
Telex No. Kar. 2832, Karachi

RECEIVED JAN 1 2 1988

To Jack Shannon
with thanks [signature]

Our Ref CPT/006/1003

Your Ref

Date December 20,1987

ROBERT J. WATSON,
DIRECTOR, Q707 SALES,
TRACOR AVIATION, INC.,
SANTA BARBARA AIRPORT,
495 SOUTH FAIRVIEW AVENUE,
SANTA BARBARA, CALIFORNIA 93117.

SUBJECT:- <u>CAA(PAK) APPROVAL FOR JS 1176 APPENDICES 10 & 11</u>

Civil Aviation (Pakistan) has granted us approval to operate the Q707 aircraft under APPENDIX 10 (Non-Noise Limited Operation) and APPENDIX 11 (ICAO Annex 16, Chapter II noise characteristics) of the document JS 1176.

A duly signed copy of document JS 1176 is enclosed here which may please be forwarded to Shannon Engineering for incorporation of the Pakistan Civil Aviation Certification in this document.

(CAPT. M. A. SALIM)
<u>CHIEF PILOT TECHNICAL</u>

/M.R./SASB/

UNITED ARAB EMIRATES
Ministry of Communications
Directorate General of Civil Aviation

P. O. Box : 900
ABU DHABI

Tel. : 362900
Telex : 22668 EM

وزارة المواصلات
الادارة العامة للطيران المدني
ص ب : 900
ابوظبي

هاتف : 362900
تلكس : 22668

Ref. : 2/2.194
Date : 10 September 1988

Dubai Air Wing,
P.O. Box 11097,
Dubai - (UAE).

Attn: **Mr. A.H. Everett**
      Engineering Director

Dear Sir,

B707 - 3L6C A6-HRM
Quiet Nacelle Modification

This is to approve the AFM Supplement Doc JS 1326 as prepared by Shannon Engineering Inc. for the additional Certificate Limitations, Procedures and Performance Information to permit operation with the Q707 Modification installed in the subject aircraft.

Specifically, this approval includes Appendix 10 to the above mentioned supplement which permits take off and landing operations of the Q707 modified aircraft at airports which do not require compliance with noise limitations of ICAO Annex 16.

Yours faithfully,

(M.J. Dobson)

/usk

XVII Community Transports Inc. AKA ComTran, Ltd.

I mentioned earlier that in order to conserve the funder's cash flow I had agreed that one half of the compensation I was to receive in the form of royalties was to come from Community Transports, Inc. in the form of a perpetual engineering services agreement, an ESA. According to that agreement, Douglas Jaffe, acting as Community Transports, Inc. could not engage in any aviation product development without SHANNON engineering, Inc. being the only source of engineering talent. That agreement was to cover all aviation projects undertaken by them.

The STC was awarded February 26, 1985. August 15, 1985 Douglas files a new corporation named Comtran International, Inc. [A] and by that November, Douglas was releasing press statements saying Comtran (Comtran International, Inc. on the letterhead) had developed and was selling the Quiet 707. [B] That was only two years after signing the ESA in November of 1983 as Community Transports, Inc. There were five such press releases including correspondence with me through 1992 and not one specifically mentioning Community Transports, Inc. A search of the Texas Secretary of State corporate records reveals there had been four Comtran's filed. One each in 1982, 1983, and 1985.[C] One was actually another Community Transports., but with Ltd. rather than Inc. for the corporate identity. The difference between the corporate registrations was achieved by adding international, hyphenation, or capitalization and Inc. or Ltd!

I was hoodwinked, Texas style. Community Transports, Inc. was not the same corporate entity as Comtran. We thought, as was intended, that Comtran was an abbreviation for Community Transports. The two had the same address, same letterhead, the same telephone number and same employees. We thought they were the same. But that was not true. The transaction was an asset transfer from Community Transports Inc to Comtran, without the liability – my ESA!

We were very busy through 1993 with subsequent and foreign 707 variants. Between 1989 and 2009 Douglas Jaffe was financially involved in fourteen aviation programs, all in violation of the ESA. But by February 14, 1990 the statue of limitations had run out on us. Lawyers in Texas and Seattle just shook their heads. Part of the fault was mine in that I had drafted the ESA. I was without recourse.

Douglas Jaffe never fulfilled his obligation.

In retrospect his lack of integrity is well known: he and two of his attorneys were indicted in 1992 for conspiring to secretly contribute funds to Democratic political campaigns in 1988. [D] Internet searches reveal that he is frequently referred to as belonging to "the Texas Mafia," and there is reference to his involvement in many shady deals. [E]

However, Doug, through his corporate alter ego, provided invaluable assistance. He provided the initial test airplane, N886PA, he sponsored the Paris and Farnborough chalets and he arranged financing for many of the sales. These two were critical given the large number of foreign airplanes in the market. Many of those airplanes were government possessions. That environment contained an inherit risk that Tracor, Inc. would necessarily avoid. Look at the preponderance of foreign customers listed in section XVI, and ask yourself how you would have protected your loan if you had assisted them in purchasing the hush kit. The airplane most certainly had to have been the collateral. But how would you have repossessed if the deal went sour? Doug's appetite for high risk investments and his experience in questionable business practice was essential in penetrating the market.

XVIII Post Flight

Success or failures are mere perceptions. The value is in the journey.

It was a long time before we realized what we had done. We increased the number of 707s operating in worldwide markets –a credit recognized by the Boeing Company. The special issue of *INC.* Magazine for December 1986 recognized us one of the fastest growing companies in the USA. It was the first time a small engineering company successfully made a modification to a Boeing airplane and the first time any modification was accepted by foreign regulatory agencies. We so dominated the market that in the end there was no competition. And, we did that against the tide of congress through granting exemptions and the attempts by a major airplane manufacturer to obstruct us.

I keep up with John Alberti, Larry Timmons, Scott Toner and Jane Weld Kaszycki. And I have visited with Galen Haws and Charlene Hinton recently. For years when I was in Dallas I would have dinner with Mike Conlin and his wife Mike. Four of the young engineers I worked with went on to senior positions in the FAA. I considered that as a compliment. They had vastly more varied experience at SHANNON engineering, Inc., than they could have possibly had anywhere else. That no doubt added to their contributions to the FAA. Four senior individuals went on to form their own separate company. Three are still in existence some twenty years later. That I also take as a compliment.

For years I would explain to people that we had an engineering company. With the perspective of time I now believe it was a business school for engineers.

It is difficult to grasp the impact it had on me personally. I had the opportunity to do things barely comprehended by the traditional engineer. I gained deep respect for salesmen. I traveled with them to foreign lands where subtle cultural and political matters would make or break a deal. I saw them tenderly weigh those considerations and close the deal. Those are insights beyond the grasp of most engineers.

The on-the-job-training I got with Jayne Weld Kaszycki and Charlene Hinton provided new management perspectives and attendant skills. I did not know an accrual system could be that sophisticated a management tool.

The years with SHANNON engineering, Inc., provided experiences that would have been incomprehensible to me only a few years earlier.

I have no regrets. This has been a remarkable journey.

References

I The Environment Leading to the Noise Rules

A. 14 Code of Federal Regulations, Aeronautics and Space, Part 36. Current version published January 1990, Office of the Federal Register, U. S. Printing Office, Washington.
B. Jayne's All the World's Aircraft, Edited by John W. R. Taylor, Jayne's Yearbooks, Paulton House, Shepherdess Walk, London N1, England.
C. "Supplemental Statement of Boeing Commercial Airplane Company, a Division of the Boeing Company. Prepared for U.S. House of Representatives, Committee on Inter-State Commerce, Subcommittee on Transportation and Commerce," June 27, 1979.
D. NASA CR-1709, "Investigation of DC-8 Nacelle Modifications to Reduce Fan-Compressor Noise in Airport Communities, Part V- Economic Implications of Retrofit" H. D. Walton, Ellis J. Gabby, G. B. Ferry, Jr., and N. L. Cleveland, Douglas Aircraft Company and National Aeronautics and Space Administration, December 1970. And NASA CR-1711, "Study and Development of Turbofan Nacelle Modifications to Minimize Fan-Compressor Noise Radiation. Volume I – Program Summary" The Boeing Company and National

Aeronautics and Space Administration, January 1971.
E. AD-787 610, "FAA JT3D Quiet Nacelle Retrofit Feasibility Program, Volume III," the Boeing Company, February 1974.
F. Federal Aviation Regulations Part 91.805. (previously Subpart E)
G. ICAO Annex 16, Volume 1, "Aircraft noise."
H. "New York Times," Monday, May 16,1983

II The Unique Nature of the Boeing 707 and DC-8

III Studies by Boeing, NASA, the FAA and McDonald Douglas

A. "FAA JT3D Quiet Nacelle Retrofit Program, Volume I-1 and -2, Lower Goal Design, Fabrication and Ground Testing, Volume II, Volume III Lower Goal Flight Testing, Economic Analysis and Summery, and Volume IV Compatibility Analysis and Design Study for DC-8 Aircraft" Boeing Company Wichita Kansas Division, June 1973 thru February 1974.
B. AD-787 610, "FAA JT3D Quiet Nacelle Retrofit Feasibility Program, Volume III," the Boeing Company, February 1974. Mr. Honsberger's affidavit is on page 254.

IV Company Background

V The First Step

A. 14 Code of Federal Regulations, Aeronautics and Space, Part 36.
B. ICAO Annex 16, Volume 1, "Aircraft Noise."
C. Port Authority of New York and New Jersey Rule 520/0-00 enacted by the Port Authority April 1982. Among this was the Interim Rule, which provided that effective January 1, 1983, the operators of subsonic jet airplanes exceeding 75,000 pounds in maximum certificated takeoff weight shall conduct airplane movements at John F. Kennedy International, Newark International and LaGuardia in compliance with noise limits consistent with FAR 36 requirements.
D. "the Position Report" October 1, 1981, Volume 1 No. 2, SHANNON engineering, Inc., Seattle WA.
E. "the Position Report" April 1, 1982, Volume II No. 2, SHANNON engineering, Inc., Seattle, WA.

VI Preliminary Configuration

A. "Operators of Boeing 707/720 Series Turbo-fan Powered Aircraft," W. M. Conlin, June 21, 1982, Dallas, Texas. His report showed 505 707's in service in the world. JS 1106 by John Alberti,

SHANNON engineering, Inc., showed a Boeing Study at 479.

B. PWA-3486, "Analytical and Experimental Studies for Predicting Noise Attenuation in Acoustically Treated Ducts for Turbofan Engines," Earnest Feder and Lee Wallace Dean III, Pratt & Whitney Aircraft Corporation, Eadt Hartford, Conn, National Aeronautics and Space Administration.

C. "JS1106, 707 Noise Abatement Program, Phase 1 (Feasibility) Final report," SHANNON engineering, Inc., John Alberti, July 16, 1982.

D. Letter from W. C. Chin, Manager, Manufacturing Inspection Branch, ANM -180S, Seattle Certification Office, to SHANNON engineering, Inc., July 14, 1983.

E. International Airline Transport Association, ATA, has provided a three letter code for every airport in the world. In the United States the letter K precedes those three letters to identify an airport from a nearby navigational aid or specific navigational point (such as a VOR, low frequency radio station or identified GPS coordinates).

F. "JS1127, Quiet 707-300 B ADV/C Airplane, Engine and Reduced Flap Landing, Performance and Operating Characteristics Certification Proposal" SHANNON engineering, Inc., Pat Palmer, November 15, 1983.

G. STC SA2063NM, Provision for Flaps 30 landing switch for GPWs and gear up warning for the Model 707-100B. July 12, 1983.

VII The Market for Venture Capital

A. Economic Recovery Act of 1981, Internal Revenue Code §41, also known as the Research & Development Tax Credit.
B. Hall, Bronwyn H., "R&D Tax Policy During the 1980s: Success or Failure?" The University of California at Berkeley, NBER, and the Hoover Institution, Stanford University. Published by the National Bureau of Economic Research. November 17, 1992.
C. Boswell, John H. and Coates, George Andrew, "Saving the General Aviation Industry: Putting Tort Reform to the Test" Southern Methodist University, School of Law, Journal of Air and Law Commerce, December 1994/January 1995.
D. Tarry, Scott E. and Truitt, Lawrence J., "Rhetoric and Reality: Tort Reform and the Uncertain Future of General Aviation," Southern Methodist University, School of Law, Journal of Air and Law Commerce, September October 1995.

VIII Seeking Research and Development Capital

A. "the Position Report" April 1, 1982, Volume II No. 2, SHANNON engineering, Inc., Seattle, WA.
B. "JS1106, 707 Noise Abatement Program, Phase I, (Feasibility) Final Report" J. Alberti. Initial release. July 2, 1982 through revision D November 17, 1982, SHANNON engineering, Inc., Seattle, WA.

IX Contract Negotiations

A. Document T83-AU-9551U, "Acoustical Assessment of Shannon B-707 Retrofit Program Proposal," William K. Conner, Acoustical Research Department, Tracor Aviation, Inc. March 11, 1983.

X Contract Signing

A. "Sale and Support Agreement" and "Royalty Agreement" October 25, 1983 between Tracor Aviation, Inc. and Malcus, Inc. DBA SHANNON engineering, Inc.
B. "Exclusive Engineering Services agreement" between Jack Shannon, President of Malcus, Inc. and Morris D. Jaffe, Jr., President of Community Transports, Inc. November 28,1983.

XI Development

A. "JS1106, 707 Noise Abatement Program, Phase 1 (Feasibility) Final Report," SHANNON engineering, Inc., John Alberti, July 16, 1982.
B. "JS1129, Quiet 707 FAR Part 25 and CAR 4b Compliance Document," SHANNON engineering, Inc., Jack Shannon, Original release January 12, 1984 through Revision N December 2, 1986.
C. "JS 1245, Test Log and tabulated Data for Performance Improvement Studies on a Quiet 707-300 Nacelle with JT3D-3B Engine," Rohr Thrust Stand, January 2-14, 1986, Colin Fender, SHANNON engineering, Inc., May 5, 1986. The report title implies tests were in January of 1986 but that is the date Rohr Industries published their final data. Actual test occurred around April 25, 1984.

XII flight Test

A. International Airline Transport Association, ATA, has provided a three letter code for every airport in the world. In the United States the letter K precedes those three letters to identify an airport from a nearby navigational aid or specific navigational point (such as a VOR, low frequency radio station or identified GPS coordinates).

B. "JS1162, Quiet 707 Modified Airplane Test Log and Manual Notes," SHANNON engineering, Inc., Pat Palmer, May 20, 1985.

XIII Certification

A. JS1129, "Quiet 707 FAR Part 23 and CAR4b Compliance Document", Jack Shannon, SHANNON engineering, Inc., Inc. July 1, 1986. And JS1146, "Quiet 707 Compliance with FAR Part 36 Noise Certification Standards, Gary Gorder, SHANNON engineering, Inc., August 6, 1985.
B. JS1146, "Quiet 707 Compliance with FAR Part 36 and ICAO Annex 16 Noise Certification Standards," SHANNON engineering, Inc., Gary Gorder, January 8, 1985.
C. FAA Supplemental Type Certificate Number 2699NM, issued to SHANNON engineering, Inc., February 22, 1985. And, subsequent STC's for variants in the 707 and engine types.
D. FAA Supplemental Type Certificate Number 3595NM, for the 707-100B issued to SHANNON engineering, Inc., September 26, 1986.
E. FAA Supplemental Type Certificate Number 4015NM, for the 720 Issued to SHANNON engineering, Inc., September 11, 1987.

XIV Relationships with Boeing

XV Production

A. "the Position Report" January 1, 1986, Volume I No. 1, Seattle, WA. SHANNON engineering, Inc.
B. "Analysis and Recommendations from a Uniform National Noise Policy Affecting Cargo Aircraft and Airport Access, Draft report," Cargo Aircraft Noise Policy Working Group, Leeper, Cambridge & Campbell, Inc, Alexandria, VA. August 15, 1988.
C. On December 12, 1988 FAA Administrator T. Allan McArtor issued a Decision and Final Order of Non Compliance and Default to the San Francisco Airports Commission overturning to prohibition of Quiet 707 operations. Some $8.5M was in play in federal grants to the airport.
D. "Tracor Aviation Program Administration, Quiet 707 Operational Experience," Tracor Aviation, Santa Barbra, CA. February 21, 1991.

XVI International Aircraft

A. "Operators of Boeing 707/720 Series Turbo-fan Powered Aircraft," W. M. Conlin, June 21, 1982, Dallas, Texas.
B. ICAO Annex 16, Volume 1, "Aircraft Noise."

C. JS1146, "Quiet 707 Compliance with FAR Part 36 and ICAO Annex 16 Noise Certification Standards," SHANNON engineering, Inc., Gary Gorder, January 8, 1985. And JS1374 "ICAO Annex 16 Noise Certification for the Quiet 707-300B ADV/C with JT3D-7."
D. Civil Aviation Authority (UK) Airworthiness Approval Note (AAN) No. 1924.
E. JS1182," CAA Quiet 707 Certification," C Fender, May 14, 1986.'

XVII Community Transports, Inc. AKA ComTran, Ltd.

A. Texas Corporate File Number 76351600 August 15, 1985.
B. A November 1985 "M" Magazine article on Douglas reported "Comtran was structured to handle the 707 modification." gives evidence of Douglas' intended role of Comtran, Inc. in the Quiet 707. In addition there are at least five independent press reports through July of 1992 reporting Comtran's involvement in the Quiet 707. Beginning about 1989 I have kept large number of invoices, some contracts, and many letter signed by Douglas showing Comtran as being "responsible" for the Quiet 707.
C. The ESA was signed by Morris Douglas Jaffe, Jr. President of Community Transports, Inc. November 28, 1983. Community Transport, Inc. forfeited

existence on February 14, 1995 (twelve days before the STC was awarded) with Patricia Pozza as registered agent (Texas file 12984700, Document 343857). Comtran International, Inc. was formed August 15, 1985 with Morris D. Jaffe, Jr. as president and Patricia Pozza as the in house attorney (file 76351600). Both addresses were/are 1770 Sky Place Blvd, San Antonio, TX 78216. I have multiple letters regarding the 707 signed by Comtran International's Douglas Jaffe as far back as September 26, 1989 with that address.
D. Indictment Called GOP Harassment, "San Antonio Light," July 9, 1992.
E. http://educationforum.ipbhost.com/index.php?showtopic=17920.

XVIII Post Flight

Jack has resided in Seattle since 1961. He is an avid sailor and hunter.

He has also published "the 58P," a pilot's instructional book on the Beechcraft Baron.

www.ingramcontent.com/pod-product-compliance
Lightning Source LLC
Chambersburg PA
CBHW061512180526
45171CB00001B/153